地质科普系列丛书

火山爷爷

畅游小沟背，童眼看火山

主编　张忠慧

黄河水利出版社
·郑州·

> **图书在版编目（CIP）数据**
>
> 火山爷爷：畅游小沟背，童眼看火山：汉英对照 / 张忠慧主编. — 郑州：黄河水利出版社，2020.8
> ISBN 978-7-5509-2771-1
>
> Ⅰ.①火… Ⅱ.①张… Ⅲ.①火山－少儿读物－汉、英 Ⅳ.①P317-49
>
> 中国版本图书馆CIP数据核字（2020）第145308号

火山爷爷——畅游小沟背，童眼看火山

策划统筹	《资源导刊》杂志社
责任监制	程 寰　王路平
责任编辑	田丽萍
责任校对	兰文峡
装帧设计	张 金　江吉洁
美术编辑	谢 萍

出 版 社：黄河水利出版社　　　　　　网址：www.yrcp.com
地址：河南省郑州市顺河路黄委会综合楼14层　邮政编码：450003
发 行 单 位：黄河水利出版社　　　　　　E-mail:hhslcbs@126.com
发行部电话：0371-66026940、66020550、66028024、66022620（传真）
承印单位：郑州新海岸电脑彩色制印有限公司
开　　本：889 mm×1194 mm　1/20
印　　张：7
字　　数：120千字　　　　　　　　　　印　　数：1—2 000
版　　次：2020年8月第1版　　　　　　　印　　次：2020年8月第1次印刷

定价：68.00元

编委会

主　　任：卢多璋

副主任：冯长春　刘晓玲

成　　员：孔会文　梁　凯　杨　森　王　芳
　　　　　卫豆豆　王严亮

主　　编：张忠慧

副主编：任利平　江吉洁

成　　员：张　金　井　燕　危红梅　渠玉冰
　　　　　赵潇予　张三衡　宋丹萍　张　睿
　　　　　石晨霞

编　　制：河南省山水地质旅游资源开发有限公司

王屋山—黛眉山世界地质公园网站网址
中文版网址：http://www.geopark-wdgy.com/
英文版网址：http://www.geopark-wdgy.com/en/

人物介绍

我们都是自然爱好者。我们的足迹已遍布了大半个中国。这不,上个周末,我们就在"路上"与17亿年的火山爷爷经历了一场奇遇……

爸爸
工程师

妈妈
教师

毛毛
四年级小学生

周六上午,毛毛像往常一样去青少年活动中心上课。一到那里,巨大的LED屏便吸引住了他。

毛毛兴奋地盯着手里的海报。突然，火山爷爷从海报上站了起来。

毛毛！爷爷带你去王屋山—黛眉山世界地质公园来一次空中体验吧。

毛毛不由自主地随着火山爷爷飞起来，坐进了一艘飞船里。

王屋山

愚公群雕

黄河

快看！那条大河就是黄河，是中华民族的母亲河；那座像大屋子一样的山是古时候黄帝祭天的地方，叫王屋山；它的脚下就是愚公移山的地方；那座像龟背一样的山就是鳌背山，山下就是我们的目的地——小沟背了。

飞船缓缓地降落在了小沟背景区入口广场上。

紧接着,火山爷爷变成了广场周围黑色的火山岩,飞船变成了下面的大古石组红色砂岩。只剩下毛毛孤零零地站在广场上。

毛毛慢慢往前走，天渐渐变黑，山风呼啸。他吓得哇哇大哭。

天空上升腾起片片金花。毛毛朝光亮处跑去。

在发光的地方，舌舌看到了从未见过的奇妙景象。空地上有个冒火的炉子，里面正熬着红彤彤的东西，一个老头舀了一勺，朝天上一洒，黑色的天空马上就金光璀璨，如天女散花一般，还发出"嗞啦嗞啦"的声音。

舌舌走近一看，这个老头长得可真奇怪，他的身体就好像很多个枕头堆在一起。

我的形成过程和这些小铁粒有点相像。17亿年前，小沟背这里是一片汪洋大海，我还是地层深处的岩浆，体温一般在上千摄氏度，像刚出炉的铁水，火红炽热。地球母亲给予我无穷力量，让我在这片大海里喷发，形成海底火山。

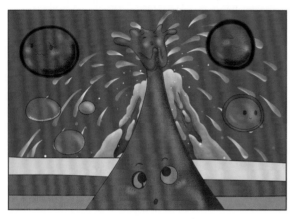

①有一天，火山在海底喷发了，大量岩浆喷薄而出。

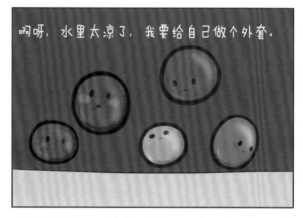

②当炙热的岩浆碰到冰冷的海水，瞬间变成一个个内软外硬的团块，像是一颗颗有弹性的"脆皮软糖"。

③这些"脆皮软糖"要比海水重多了，慢慢沉到了海底，相互挤压在一起。

④这就是我们枕头部落居民的形成过程。

地质学家在小沟背发现了我们，由此推断这里曾经有海底火山存在。

眼前出现一座像花卷馍一样的小房子。一位胖胖的、笑眯眯的老奶奶站在门口,她的身上布满一圈一圈的花纹。

欢迎来到圈圈饭店!

屋子里到处都装饰着不规则的圈圈。桌子上已经为毛毛准备好了面条和花卷。

奶奶，你的身上为什么到处都是圈圈？是不是和枕头爷爷一样都是火山里的岩浆变的，它和我们的花卷、面条之间有关系吗？

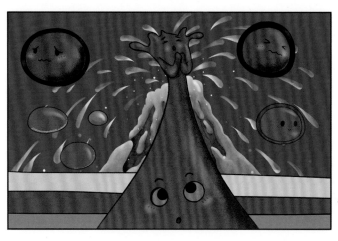

①我和枕头爷爷一样，老家也在海底的火山里。我们的年纪差不多。

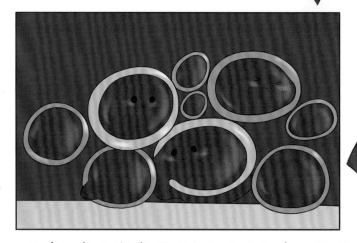

②喷出来的岩浆球球在海底堆起来，下面的一些球球不堪重负，外壳破裂，把里面的岩浆挤出来。

④随着新球球不断落下，压力不断加大，球球表面不断破裂，内部的岩浆不断被挤出。如此循环往复，形成了一圈圈的结构。

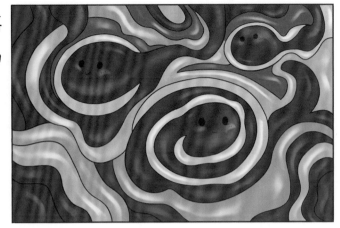

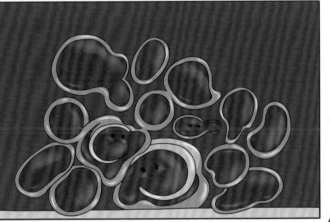

③被挤出的岩浆覆盖在被挤破的外壳表面。

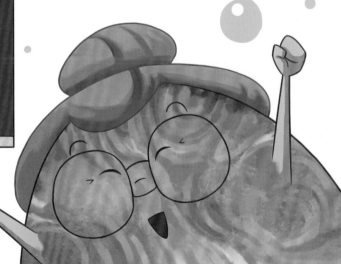

这就是我的形成过程。

毛毛正准备吃饭，听到外边一阵阵叫好声。隔着窗户一看，饭店门前的广场上一个浑身是洞的老爷爷正在吹泡泡，周围围了好多人。

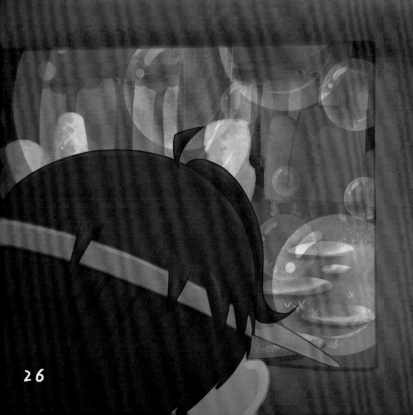
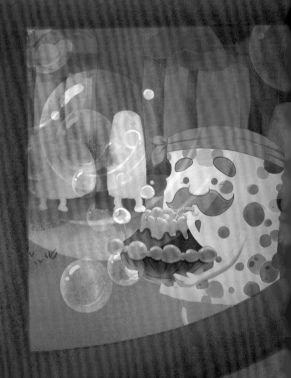

过了一会儿,那位表演吹泡泡的老爷爷走进了圈圈饭店,吵着要吃饭。

这位爷爷接过碗就狼吞虎咽地吃起来。可是面条从他的嘴巴进去没多久,就从他身上的洞洞流了出去,像洒水壶一样。

吃完饭，这位爷爷自觉搞好了卫生，开开心心地走了。

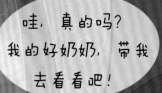

哇，真的吗？我的好奶奶，带我去看看吧！

对了，洞洞爷爷还有几个长得很好看的妹妹呢，她们都住在"花花部落"。

于是，圈圈奶奶就带着毛毛来到了一个大广场上。这里有唱歌的，有跳舞的，还有画画的，真热闹啊！

很久以前她们身上也有很多个窟窿，后来她们在补洞洞工作室对洞洞进行了修补，有的补得像杏仁，有的像梅花，有的像荷花。

天慢慢亮了,太阳也快要升起来了。毛毛看到远处出现了一片雄伟壮观的石头柱子。

柱子部落

上面有柱子爷爷在等你,我就带你到这里吧。

谢谢爷爷!再见!

想当年我从火山口里喷出来的时候,也还是一大片稠糊糊的岩浆,把周围的地面都给覆盖上了,占据了好大的地盘。

可是没过多久，随着身体慢慢变冷，体积收缩，身体表面出现很多像龟背一样的裂纹，这些裂纹甚至切穿整个熔岩层，形成一根根像铅笔一样的柱子。

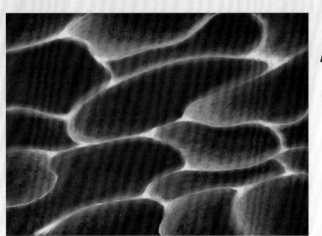

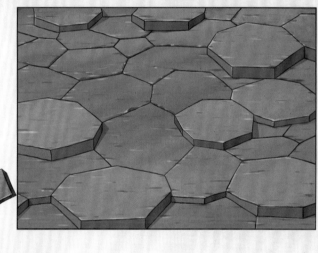

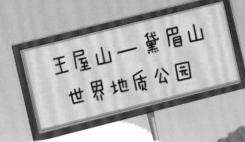

爷爷，您居住的小沟背已成为王屋山一黛眉山世界地质公园的一部分了。

那就好，那就好。

这里的所有地质遗迹都会得到最大程度的保护。人们来到这里主要是学习地球知识、欣赏奇妙景观的。

毛毛继续往前走，眼前又出现了带他来到小沟背的火山爷爷。

毛毛今天玩得开心吗？

开心极了！今天我认识了好多好多有趣的爷爷奶奶。

哈哈，今天你所见到的爷爷奶奶们其实都是17亿年前火山所留下的各种神奇现象。它们都是极为珍贵的地质遗迹。

枕头部落

圈圈部落

小沟背火山遗迹线路图

毛毛睁开眼睛，原来是做了一个梦。他赶紧跳下车，迫不及待地要到景区看一看梦里的爷爷奶奶究竟长什么样子。

如果你想知道更多关于17亿年前的火山秘密,欢迎来济源小沟背,这里有讲述不尽的火山爷爷和奶奶们的古老故事。

火山家族人物介绍

昵称：枕头爷爷

学名：枕状熔岩

出生地：海底火山

所属家族：岩浆岩——火山岩

 生平简介

岩浆在水下喷发或在陆地喷发落入水中，由于温度急剧变化，会形成一个个岩浆球。这些岩浆球表面接触冷水首先会形成硬壳，落入水底后，堆积在一起的岩浆球受水压和岩浆球本身的压力，它们又会变成椭球形，像一个个"枕头"排列在一起。这种具有"枕头"状构造的火山岩就叫做"枕状熔岩"。

昵称：圈圈奶奶

学名：淬碎熔岩

出生地：海底火山

所属家族：岩浆岩——火山岩

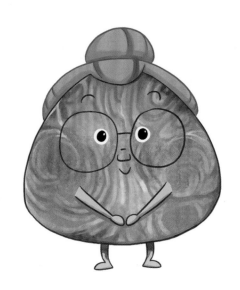

 **生平简介**

　　火山喷发的炽热岩浆冲入水中变成一个个岩浆球，岩浆球表面遇水迅速冷却形成硬壳。沉入水底后，岩浆球的外壳又由于冷却收缩及内部压力作用而发生破裂，导致球内的岩浆重新溢出冷却。一次次的破裂和溢出，形成环状构造，具有这类构造的岩石称之为"淬碎熔岩"。

昵称：洞洞爷爷

学名：气孔状安山岩

出生地：陆地火山

所属家族：岩浆岩——火山岩

生平简介

气孔状安山岩是一种喷出岩，在岩浆喷发的过程中会携带大量的气体。如果气体没有及时逸出，在成岩过程中就会被包裹在岩石中间。受地壳运动的影响，岩石中会发育大量的节理裂隙，包裹在岩石中的气体就会顺着节理裂隙逸出，留下气孔。当气孔充分发育时，岩石会变得很轻，甚至可以漂在水面，形成浮岩。

昵称：柱子爷爷

学名：玄武岩柱状节理

出生地：陆地火山

所属家族：岩浆岩——火山岩

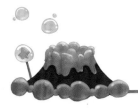 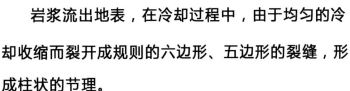 生平简介

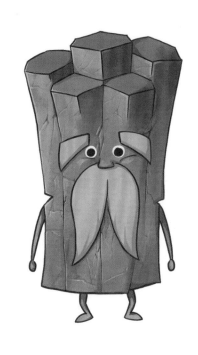

　　岩浆流出地表，在冷却过程中，由于均匀的冷却收缩而裂开成规则的六边形、五边形的裂缝，形成柱状的节理。

昵称：杏仁奶奶

学名：杏仁状安山岩

出生地：陆地火山

所属家族：岩浆岩——火山岩

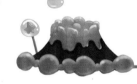 **生平简介**

对于含有大量气孔的火山岩，在地壳运动过程中，气孔内填充了大量富含矿物质的溶液。溶液中的矿物质在气孔中沉淀下来，就会在暗色岩石上留下白色的斑块，形似杏仁，故名杏仁状安山岩。

昵称：梅花奶奶

学名：杏仁状安山岩

出生地：陆地火山

所属家族：岩浆岩——火山岩

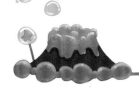 ### 生平简介

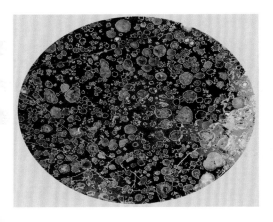

对于含有大量气孔的火山岩，在地壳运动过程中，气孔内填充了大量富含矿物质的溶液。溶液中的矿物质在气孔中沉淀下来，就会在暗色岩石上留下其他颜色的斑块，形似梅花。气孔间常有曲折的细裂隙被矿物填充，形成了"梅花"的枝干。

Nickname : Grandma Plum Blossom

Scientific Name : Amygdaloidal Andesite

Birthplace : Land Volcano

Family Attribution : Magmatic Rock-
Volcanic Rock

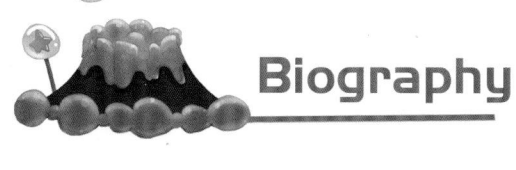

Biography

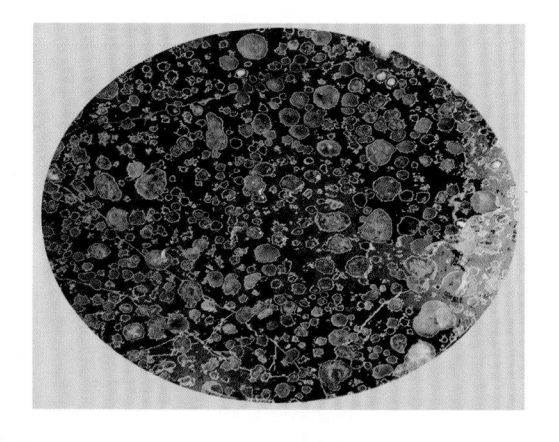

As to volcanic rocks with a large number of stomas, in the process of crustal movement, these stomas will have a mass of rich mineral solutions. The minerals in the solution precipitated in the stomas, leaving patches with different colors in the dark rocks, shaped like "plum blossom". The zigzag fine fissures among the stomas are often filled with minerals, forming the branches of the plum trees.

Nickname : Grandma Amygdale

Scientific Name : Amygdaloidal Andesite

Birthplace : Land Volcano

Family Attribution : Magmatic Rock-
　　　　　　　　　Volcanic Rock

Biography

As to volcanic rocks with a large number of stomas, in the process of crustal movement, these stomas will have a mass of rich mineral solutions. The minerals in the solution precipitated in the stomas, leaving white patches in the dark rocks, shaped like almonds, hence it named amygdaloidal andesite.

Nickname : Grandpa Pillar

Scientific Name : Basalt Columnar Joint

Birthplace : Land Volcano

Family Attribution : Magmatic Rock - Volcanic Rock

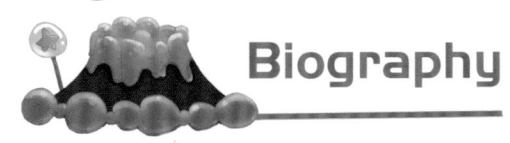

Biography

When magma flows out of the surface, during the process of cooling, it splits into regular hexagonal and pentagonal cracks as a result of uniform cooling contraction, which constituted the columnar joints.

Nickname : Grandpa Dongdong

Scientific Name : Stomatic Andesite

Birthplace : Land Volcano

Family Attribution : Magmatic Rock -
Volcanic Rock

Biography

Stomatic Andesite is an extrusive rock that carries a lot of gas as magma erupts. If the gas does not escape immediately, it will be wrapped in the rock during diagenesis. Under the influence of crustal movement, a large number of joints and fissures will develop in the rock, and the gas wrapped in the rock will escape along the joints and fissures, leaving stomas. When the stomas are well developed, the rock becomes so light that it floats on the water, forming pumice.

Nickname : Grandma Quanquan

Scientific Name : Quenched Lavas

Birthplace : Submarine Volcano

Family Attribution : Magmatic Rock -
Volcanic Rock

Biography

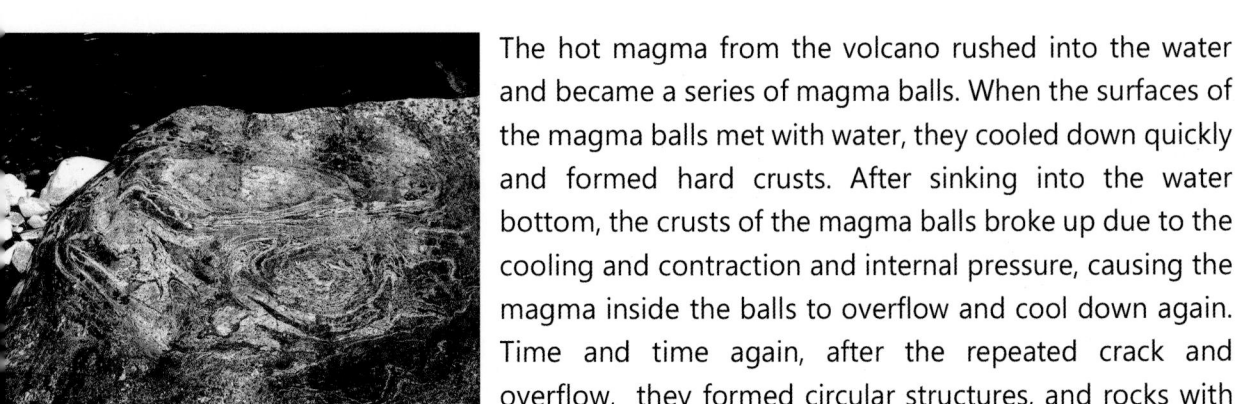

The hot magma from the volcano rushed into the water and became a series of magma balls. When the surfaces of the magma balls met with water, they cooled down quickly and formed hard crusts. After sinking into the water bottom, the crusts of the magma balls broke up due to the cooling and contraction and internal pressure, causing the magma inside the balls to overflow and cool down again. Time and time again, after the repeated crack and overflow, they formed circular structures, and rocks with such structures are called "Quenched Lavas."

Characters Introduction

Nickname : Grandpa Pillow

Scientific Name : Pillow Lava

Birthplace : Submarine Volcano

Family Attribution : Magmatic Rock -
 Volcanic Rock

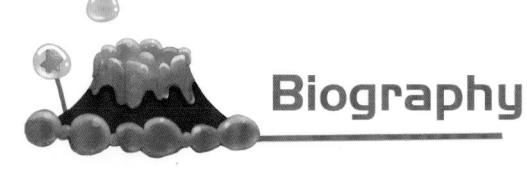 Biography

Magma erupted underwater or fell into the water after the land eruption, and formed a train of magma balls because of the sharp change in temperature.The surfaces of these magma balls first form hard crusts when they contact with cold water, but the pressure from the water and the magma balls themselves forced them to form ellipsoids when they fall to the water bottom, arranged together like pillows. This type of volcanic rock with a pillow-like structure is called "Pillow Lava".

If you want to know more about the secrets of volcanoes formed 1.7 billion years ago, welcome to Xiaogoubei in Jiyuan City, where there are countless old stories of grandpa and grandma volcanoes.

Diudiu opened his eyes and realized that he just had a dream.
He jumped out of the car and couldn't wait to confirm what the
grandparents in the dream looked like.

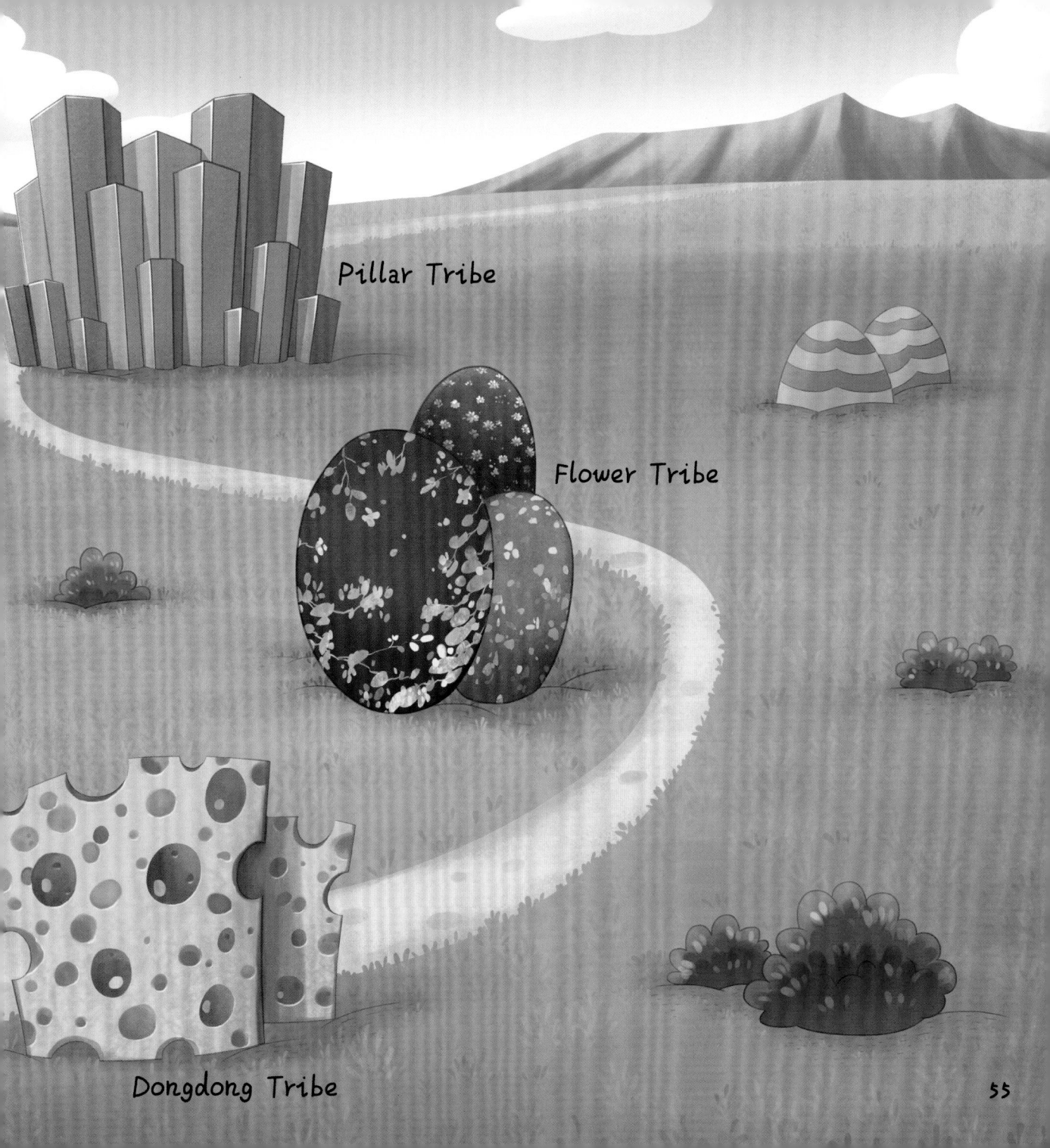

Pillar Tribe

Flower Tribe

Dongdong Tribe

Diudiu walked on, Grandpa Volcano who took him to Xiaogoubei reappeared.

Did you have a good time today, Diudiu?

So happy! Today I met many interesting elders.

Haha, the grandparents you have seen today are actually the magical phenomena left behind by the volcano formed 1.7 billion years ago. They are extremely precious geoheritages.

Each pillar here is a precious geoheritage, and my task now is to protect them. So I have to check the number of pillars every day.

Ah, I got it, the scientific name of these pillars is basalt columnar. I've seen it in Northern Ireland and Hong Kong Global Geopark, China. To tell the truth, you are much cooler than Grandpa Pillow and Grandpa Dongdong! Then why do you count?

But soon, as the body got cold and shrank, cracks occurred on the surface like a turtle's back, and these cracks even cut through the entire lava layer to form pencil-like pillars.

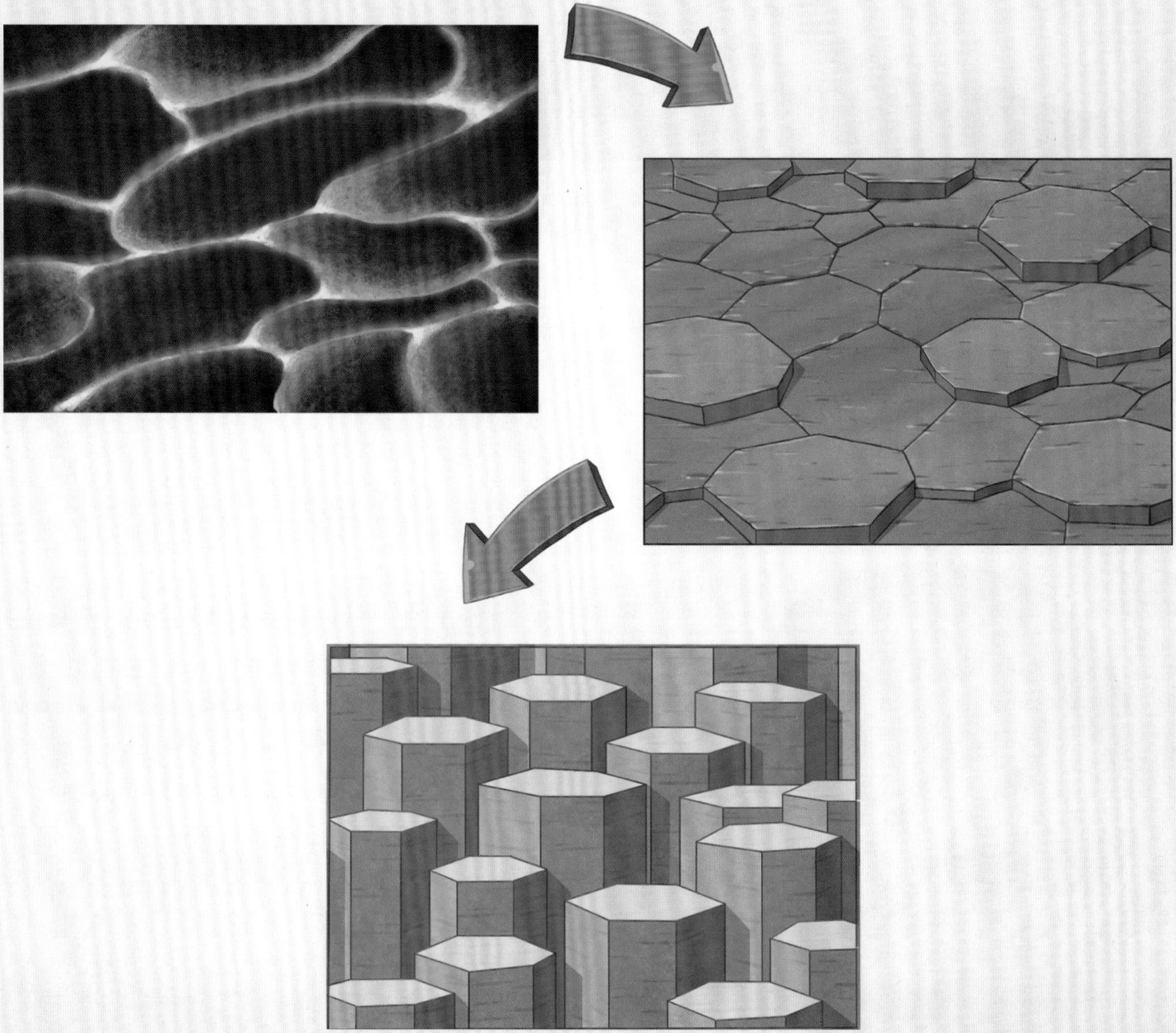

When I erupted out of the crater, there was still a mass of thick, sticky lava that covered the ground around me and took up a lot of space.

Diudiu went to the top of the mountain, and saw Grandpa Pillar who was keeping on mumbling and counting.

Hello, Grandpa Pillar!
What are you counting?

It was getting light and the sun was about to rise. Diudiu saw a block of spectacular stone pillars in the distance.

Pillar Tribe

Grandpa Pillar is waiting for you upstairs. I just take you here.

Thanks, grandpa! Good bye!

Mending Studio

1000   500

It's so hot!

Holes mending materials

They also had many holes in their bodies long ago,
and then they repaired them in the Mending Studio,
some like almonds, some like plum blossoms,
and some like lotus flowers.

How beautiful they are! How could they be Grandpa Dongdong's sisters?

So the Grandma Quanquan took Diudiu to a big square. On the square, some were singing, some were dancing and some were drawing. It was quite lively!

As the magma rapidly solidified into rocks, a small amount of gas that did not escape remained. Later, numerous fissures appeared in the rocks as the crustal movement. The gas ran out of the fissure, leaving a lot of holes. This is how Grandpa Dongdong was formed.

Since then, he fell in love with blowing bubbles.

In those days, Grandpa Dongdong was still magma in the volcano, and there was plenty of gas in the magma. After erupting out to the surface, a lot of gas overflowed from the magma like blowing bubbles.

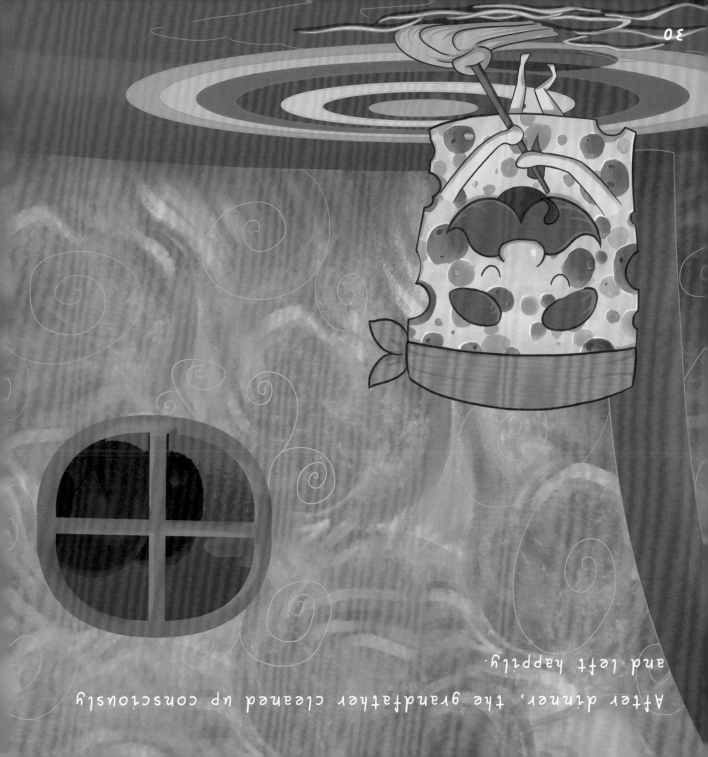

After dinner, the grandfather cleaned up consciously and left happily.

The grandfather took a bowl and began to devour it. But it didn't take long for the noodles to go in his mouth, they ran out of the holes in his body like a kettle.

The room was decorated with irregular circles everywhere. Noodles and steamed rolls had been prepared for Diudiu on the table.

Grandma, why do you have circles all over your body? Do you form from magma in the volcano as same as Grandpa Pillow? Does it have anything to do with the steamed rolls and noodles we eat?

① Like Grandpa Pillow, our home is in a volcano under the sea and we are almost the same age.

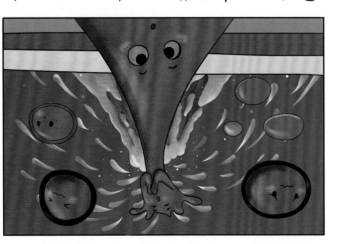

② The erupted magma balls were piled up on the bottom of the sea, and some of the balls below became overwhelmed that their crusts cracked, forcing the magma out of them.

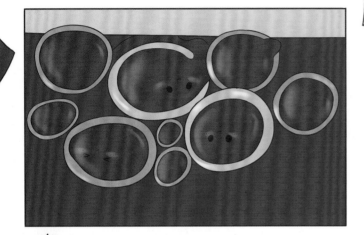

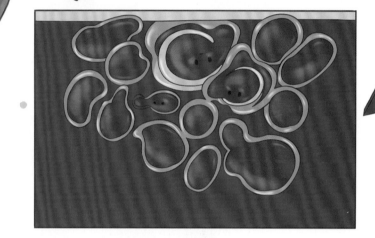

③ The extruded magma covered the surface of the cracked crust.

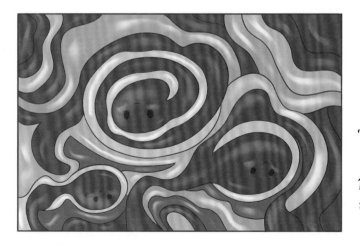

④ As the new balls continued to fall, the pressure increased, the surface of the ball continued to break and the magma inside was being pushed out. This cycle repeated, forming the circular structures.

This is how I came to be.

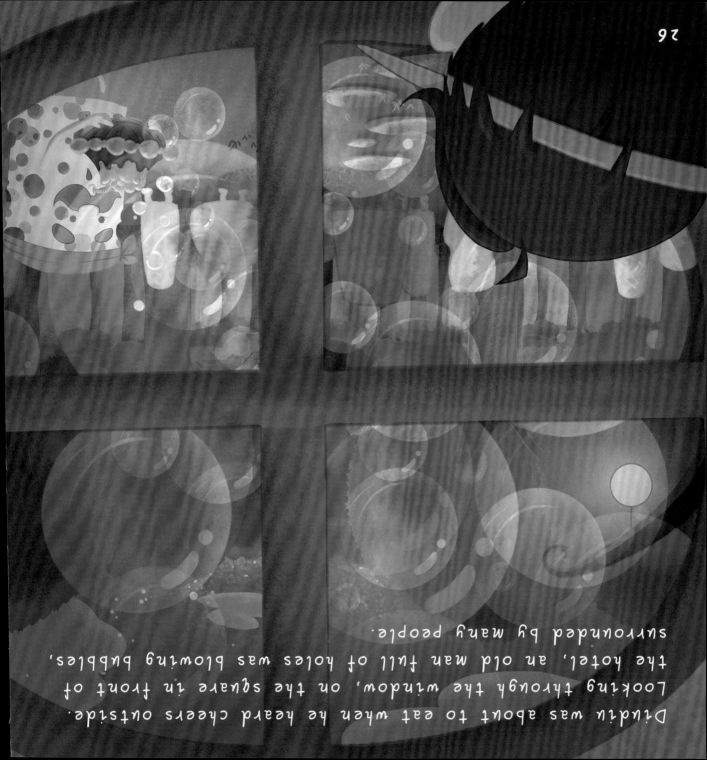

Diudiu was about to eat when he heard cheers outside. Looking through the window, on the square in front of the hotel, an old man full of holes was blowing bubbles, surrounded by many people.

"Welcome to the Quanquan Restaurant."

A small house like steamed rolls came in sight. A fat, smiling old woman stood in the doorway, whose body was covered with circles of patterns.

Geologists found us at Xiaogoubei and inferred that
there had been an submarine volcano.

The process of my formation is a little like this. 1.7 billion years ago, Xiaogoubei was a vast ocean, and I was still deep underground magma with thousands degree temperature, which looked like the hot molten iron, red and hot. The Mother Earth powered me to erupt and formed the underwater volcano in this ocean.

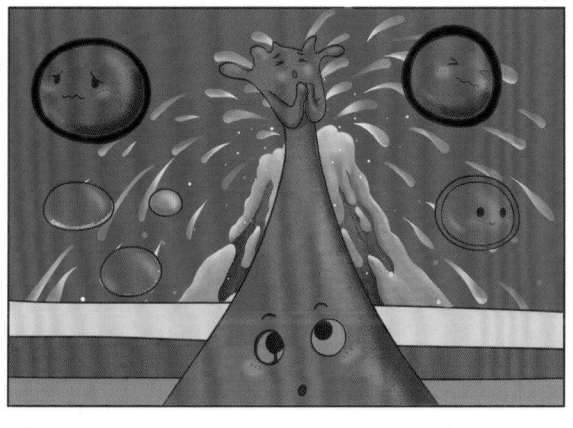

① One day, the volcano erupted at the bottom of the ocean, and a great deal of magma burst out.

② When the hot lava met the cold water, it turned into clumps, which were soft inside and hard outside, like the elastic "crispy jellies".

③ These "crispy jellies" were much heavier than the seawater, and they slowly fell to the bottom and squeezed together.

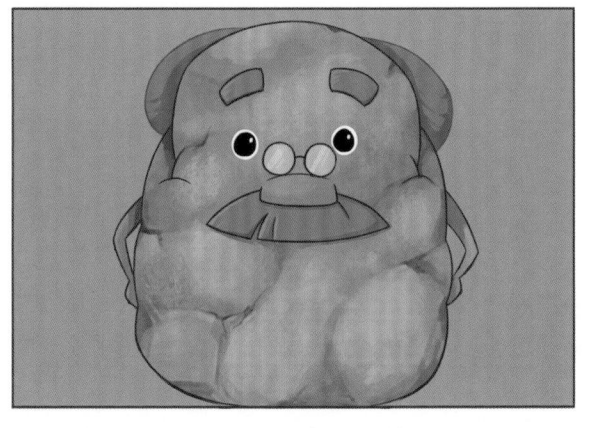

④ This is how our residents of Pillow Tribe came to be.

Diudiu saw a wonderful sight, he had never seen before in the place where the light shines. There was a burning stove in the open area where something red was boiling. An old man scooped up a spoon and poured it into the air. The black sky became golden at once, as if fairy is scattering flowers, and made a "sizzle sizzle" sound.

Pillow Tribe

Diudiu took a closer look at the old man, and he looked really strange—his body was like pillows heaped together.

Patches of golden flowers rose in the sky.
Diudiu ran to the light.

Diudiu slowly moved forward and the weather began to get dark, varying with mountain winds. He was scared to cry.

Gee? What's that noise?

Sizzling

Grandpa Volcano transformed into black volcanic rock around the square and the spaceship transformed into the red sandstone of the Dagushi Formation below. Diudiu was left standing alone on the square.

The spaceship landed slowly on the Entrance Square of Xiaogoubei Scenic Area.

Wangwu Mountain

Yugong Statues

Yellow River

Look! The big river is the Yellow River, and that is the mother river of the Chinese nation; The mountain that looks like a big house is the place where Emperor Huang worshipped the heaven in ancient times, named Wangwu Mountain; At its feet it is the place where Yugong moved the mountains; The mountain that looks like turtle's back is the Aobei Mountain, down the mountain is our destination — Xiaogoubei.

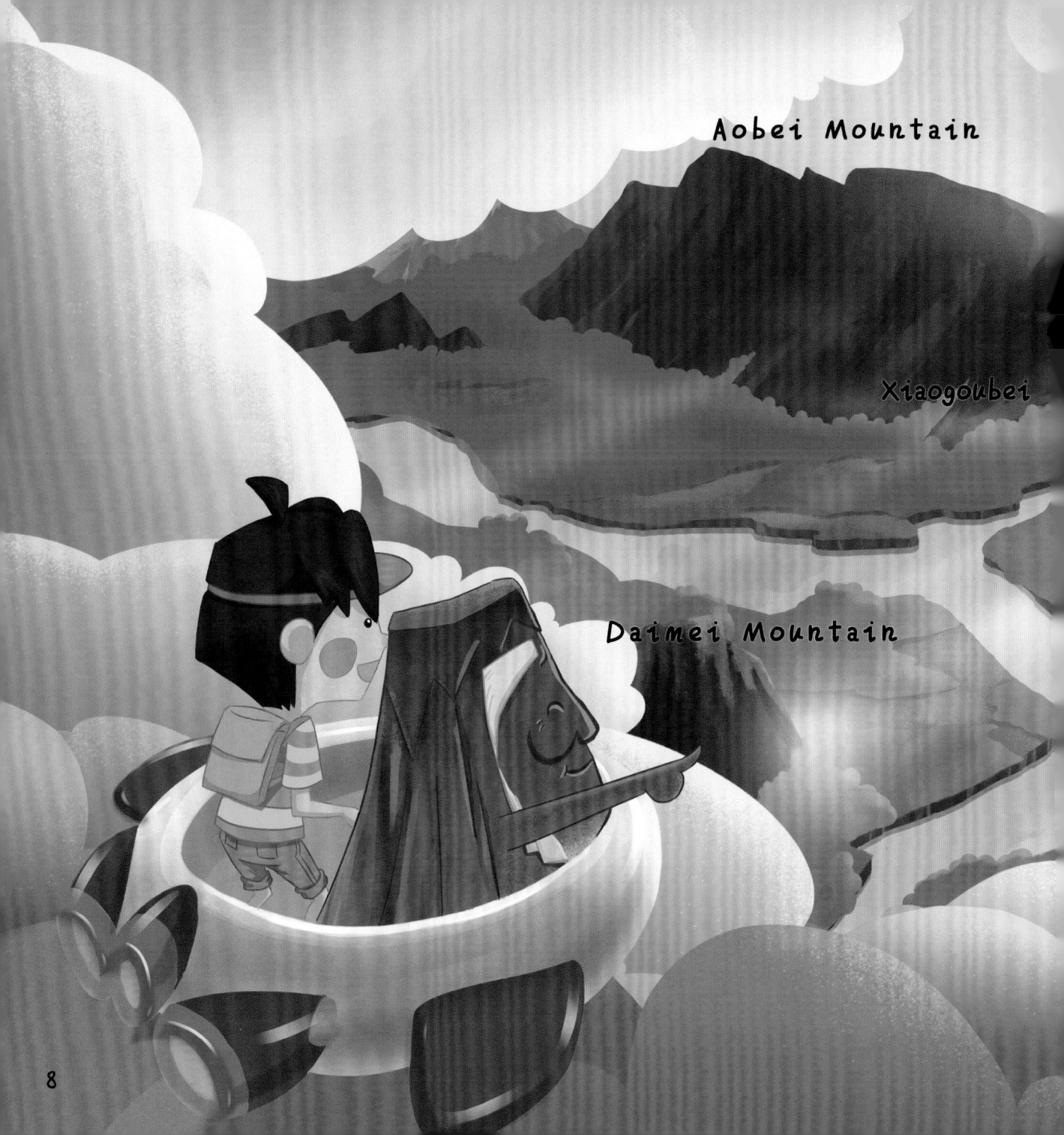

Aobei Mountain

Xiaogoubei

Daimei Mountain

Diudiu involuntarily flew with Grandpa Volcano and got into a spaceship.

When Diudiu stared at the poster in hand excitedly, Grandpa Volcano was standing up from the poster suddenly.

Diudiu, how about having an air experience with Grandpa?

Grandpa Volcano

On Saturday morning, Diudiu went to the Youth Activity Center for classes as usual. The huge LED screen caught his eyes as soon as he got there.

I am 1.7 billion years old.

Wow, it's so old! But what does it look like?

Volcano Age Competition

Jiyuan Xiaogoubei Volcano
about 1.7 billion years old
Yandangshan Volcano
about 120 million years old
Changbaishan Volcano
about 2.8 million years old
Wudalianchi Volcano
about 700 thousand years old

Youth Activity Center

Characters Introduction

We are all nature enthusiasts and our footprints have spread over half of China. Well, last weekend, we just had an encounter with 1.7 billion years old Grandpa Volcano on the way.

Diudiu
Pupil in Grade Four

Dad
Engineer

Mum
Teacher

Editorial Committee

Director: Lu Duozhang

Vice directors: Feng Changchun   Liu Xiaoling

Committee members: Kong Huiwen   Liang Kai   Yang Sen   Wang Fang   Wei Doudou
Wang Yanliang

Editor-in-chief：Zhang Zhonghui

Vice editors-in-chief：Ren Liping   Jiang Jijie

Committee members：Zhang Jin   Jing Yan   Wei Hongmei   Qu Yubing   Zhao Xiaoyu
Zhang Sanheng   Song Danping   Zhang Rui   Shi Chenxia

Compilation:Henan Shanshui Geological Tourism Resources Development Co., Ltd

Website address of Wangwushan-Daimeishan Global Geopark
Chinese Website: http://www.geopark-wdgy.com/
English Website: http://www.geopark-wdgy.com/en/

Series of Geological Popular Science

Grandpa Volcano

Exploring Xiaogoubei, and watching the volcano from children's eyes

Editor-in-Chief Zhang Zhonghui

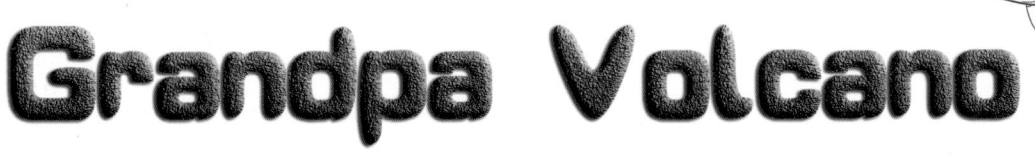

THE YELLOW RIVER WATER CONSERVANCY PRESS

· Zhengzhou ·